The Importance of Otto Gottlieb for the Chemistry of Natural Products and Brazilian Biodiversity

Marcos Aurélio Gomes da Silva

ELIVA PRESS

ELIVA PRESS

Marcos Aurélio Gomes da Silva

In this book I seek to clearly address the importance of Otto Gottlieb for biodiversity in Brazil and bring to light some concepts of the chemistry of natural products and their applications in other areas of knowledge. I do a survey of the Brazilian biomes and the importance of their conservation for the species. I reflect on the importance of the chemistry of natural products for contemporary society in order to bring solutions for the best use of our available natural resources

Published by Eliva Press SRL
Address: MD-2060, bd.Cuza-Voda, 1/4, of. 21 Chişinău, Republica
Moldova
Email: info@elivapress.com
Website: www.elivapress.com

ISBN: 978-1-63648-092-3

Contents

Otto Gottlieb considered that, at the end of the last century, understanding the mechanisms of nature was of essential importance for the future of life on planet Earth and constituted the most significant challenge for scientists living on the last biological frontiers. This interdisciplinary approach permeated his entire work, but concerned only with a theme that was the preservation of biodiversity in Brazil.

The research developed by Otto Gottlieb revealed a certain ``language`` of the plants found in the Amazon, revealing a true interdisciplinary synthesis on plant chemistry and his research on the evolution of botany in Brazil which opened new avenues for this science in the country.

Otto Gottlieb coordinated the USP natural products laboratory, and in this laboratory he coordinated research on plant chemistry with pharmacological application and was also the creator of a discipline and research area called chemosystem, a discipline that aims to quantify and map the biodiversity, helping to indicate the value of the chemical and pharmacological potential of its plant species in a region; To this end, it broadens the parameters traditionally used, not limited to external aspects of plants such as leaves and flowers, and, above all, integrates chemistry with biology, ecology and geography.

Neolignans as previously mentioned and lignans were the subject of a vast study by Otto Gottlieb and their pharmacological properties. Neolignans and lignans comprise a large class of natural products with a high diversity in their chemical structure and pharmacological activities, consisting of two phenyl propanoid units.

Neolignans, it was later discovered, also have anti-inflammatory properties. Other scientists have also proven its effects on altering the diuresis of the transmitting agent of Chagas disease.

Regarding its scientific production, there were more than 650 publications, including articles, book chapters, patents, 652 invited conferences (466 in

Brazil and 186 in 26 other countries), 145 courses taught (93 at postgraduate level).

In an article published in the magazine Natural Products Letters in 1993, co-authored with Maria Auxiliadora Kaplan of UFRJ, Gottlieb proposed the redox theory, by which species,they must adapt to antagonistic processes of fighting for and against water and oxygen to survive. That is, both water and oxygen can be either beneficial or harmful if they are in excess or lacking.

27 years ago, Gottlieb and Mors emphasized the urgency of the awakening of Brazilian scientists to a more applied research on natural products and their etiotropic properties (activity against disease-causing organisms). They envisaged more practical applications as a result of those phytochemical investigations, giving examples of the most studied national plant families with great biological potential. Currently, the search for the healing properties of plants based on their isolated constituents, "active ingredients", translates a reductionist view with regard to scientific research on medicinal plants, and this same view can be extended to other kingdoms studied from the same perspective . It is necessary to develop an interdisciplinarity involving all knowledge of popular and scientific culture, with cooperation established between groups, to establish a theoretical framework as a foundation for the production of scientific knowledge, observing its possible reflexes in the way of man to know, deal with and appropriate from nature35 and, fundamentally, reverse the acquisition of knowledge in usable technology.

Otto Gottlieb used basic biogeochemical philosophy to create new concepts, "such as the redox theory in place of the defense theory, interdependence of macro and micromolecular metabolism, and metabolism function in cell signaling".

Otto explains the system of organic nomenclature he developed: "Anyone who isolates and describes a large number of natural substances from habitually complex structures can be seen with the true seven-headed animal, the nomenclature. The rules are so varied and complicated that not even consulting a book guarantees the correct wording. With Maria Auxiliadora Kaplan, I elaborated a new system of nomenclature, in which an organic three-dimensional structure is figured as a geometric shape composed of nodes (representing atoms) and bonds. The chemical nature and relative position of the nodes are described by a numerical system. There is no need to memorize empirical terms and just know the rudiments of chemistry to be able to apply the system to even the most complex structure imaginable. "

Chemistry, under his care, led to a broad view of science. "Studying plants has allowed us to understand the rules that underpin the functioning of life, of Nature," said the scientist. Investigating the origin of life through the study of plants may seem at first glance curious. But it is exactly this correlation that forms the basis of the scientist's research on evolution, systematics and molecular ecology of plants.

Brazilian biodiversity.

Biodiversity is the variability of species of fauna, flora, microorganisms and ecosystems in a given location. It therefore consists of the genetic variety of living organisms, species of fauna and flora, habitats, ecosystems, among other elements that make up life. Brazil is the country that hosts the largest biodiversity in the world, it is estimated that in national territory are 10% to 15% of all the biodiversity on the planet. With more than 50 thousand species of trees and shrubs, it occupies the first place in plant biodiversity. No other country has so many varieties of orchids and palm trees cataloged. Its flora is composed of approximately 50 thousand different species, being that: the mammals total 530; amphibians, 517; birds, 1,677 species; reptiles total 468; plus 1.5 million insects. However, these data may be even greater, as there are thousands of species not yet cataloged in the country. Among the main factors responsible for this biological variety in Brazil are its territorial extension and the different climates. The Brazilian territory is the fifth largest in the world, its area is 8,514,876 Km2, it has large extensions of tropical forests (places that present great biodiversity) such as the Amazon rainforest, the Atlantic forest and the Pantanal. There are six types of climate present in Brazil: Equatorial, Tropical, Tropical Altitude, Tropical Atlantic, Subtropical and Semi-arid. This differentiation allows for the adaptation of different species, in addition to a great ecological diversity in the national territory. However, Brazil's biodiversity is being severely damaged by economic activities. Deforestation is one of the main villains, many species die during fires, in addition to being expelled from their natural habitats. Another aggressive practice for biodiversity is biopiracy, significantly reducing fauna and flora, causing an imbalance in the food chain.

CHEMICAL BIODIVERSITY in Brazil

For the general public, biodiversity is related to plants and animals. For a smaller contingent of the population, microorganisms also enter. However, the richness of biodiversity, likewise reveals itself in the substances, that is, in the molecules, which with amazing diversity, permeate all this natural heritage. Many of the molecules used today in the pharmaceutical and cosmetics industry are synthetic. In drugs, around 80% the origin is synthetic. Aspirin, one of the best-selling drugs in the world, is completely synthetic. Atorvastatin, used for cholesterol control and which stands out as a sales success, is a synthetic substance developed from a natural molecule. This molecular potential is another great prospect for the development from the Brazilian biomes. In this respect it is interesting to mention the Characterization, Conservation, Recovery and Use Research

5

Program Sustainable Development of the State of São Paulo, better known as BIOTAFAPESP, started in 1999.

The importance of biodiversity for the dynamics of ecological and social relations between species, including human nodes, assumes increasing proportions.The questions and discussions on such an important topic should guide not only the universe of scientists and researchers who study the incredibly fascinating richness of biodiversity but also be a call for a positioning of all social segments on what biodiversity represents for the interaction of organisms - plant / plant, animals / animals, microorganisms / microorganisms, plant / microorganisms / animals, in a perfect harmony that we call Biota, in climate changes, rainfall, desertification, etc.

Brazil, having one of the richest biodiversities in the world, has great potential to become a world reference for research on organisms from tropical environments, still with many gaps in studies on the various types of biomes, and especially on the areas of interfaces of various types of ecosystems, known as ecotones, in which there is an explosion of biodiversity and, therefore, of secondary metabolism substances - natural products. Natural products are important for the balance of nature, since regulation, adaptation and protection of the organisms that make up biodiversity are part of the fantastic vital molecular laboratory. The products originating from our natural wealth are also substances of high added value, which are currently on the rise, given the focus of large markets on bioeconomics. Thus, biodiversity, in a world shaped by the bioeconomy, has value beyond that already predicted by scientific knowledge, due to the enormous market potential. In this way business groups from central and developing countries (L'Oréal, Symrise, Bayer Health Care, Aldivia, Veleda, Cooperasión Alemana, GIZ, Developing Bank of Latino America, United Nations - UNCTAD, EXPANSCIENCE) have been investing in public opinion polls, since 2018. Union for Ethical BioTrade - UEBT is a non-profit company

6

that has been conducting research around the world on the level of awareness and perception of society about biodiversity and about the ethical supply of bioproducts in 16 countries. Since the beginning of the research 8 years ago, around 60,000 people and hundreds of companies have been asked about many important aspects related to biodiversity and the bioeconomy. Obviously, natural products are in countless radical and incremental innovations and boost a million-dollar economy in the central countries, especially those of the European community.

Brazilian biomes

The Brazilian territory, with approximately 8.5 million square kilometers, has a wide variety of natural characteristics (soil, relief, vegetation and fauna), which interact with each other forming a unique natural composition. Among the main natural characteristics that present the most variation, are biomes, sets of ecosystems with similar characteristics arranged in the same region and that historically have been influenced by the same formation processes. According to the IBGE, the country has six large biomes, which together have one of the largest biodiversities on the planet. Are they: Amazon: The Brazilian Equatorial Forest occupies about half of Brazil's territory and is concentrated in the North and part of the Midwest region. This biome is highly influenced by the equatorial climate, which is characterized by low thermal amplitude and high humidity, resulting from the evapotranspiration of rivers and trees. Its flora consists of a very rich and dense forest vegetation and presents species of different sizes - some can reach up to 50 meters in height - with large and large leaves, which do not fall in autumn. The fauna is also very diverse, composed of insects, which are present in all strata of the forest, an infinity of species of birds, monkeys, tortoises, tapirs, pacas, jaguars and others.

Cerrado: The Cerrado, or the Brazilian Savannah, extends over a large part of the Midwest, Northeast and Southeast regions of the country. It is a biome characteristic of the continental tropical climate, which, due to the occurrence of two well-defined seasons - one wet (summer) and the other dry (winter) -, has vegetation with small trees and shrubs, twisted trunks, thick bark and generally deciduous (leaves fall in autumn). The fauna of the region is quite rich, consisting of capybaras, maned wolves, anteaters, tapirs, seriemas etc.

Atlantic Forest: The specimen of Tropical Forest in Brazil has practically disappeared, because, as it was located on the country's coastal strip, much of its original vegetation was devastated to give way to the intense occupation of the coast. Originally, the vegetation of this biome was located in an extensive area of the Brazilian coast, which extended from Piauí to Rio Grande do Sul, and was constituted by a dense forest vegetation, with practically the same characteristics of the Amazon Forest: with different sizes, broadleaf (broad and large leaves) and perennial (leaves that do not fall). The fauna of this region was practically extinct and was made up of lion tamarins, otters, jaguars, giant armadillos, blue macaws and others.

Caatinga: extends throughout the Brazilian hinterland, occupying about 11% of the national territory. It is the driest region in the country, located in the semi-arid tropical climate zone. The vegetation of this region is mainly composed of xerophilous plants (accustomed to aridity, such as cacti) and deciduous plants (which lose their leaves during the driest period), in addition to some trees with very large roots that can capture water from the water table at great depths and, therefore, do not lose their leaves, like the juazeiro. The fauna of this biome is composed of a wide variety of reptiles, cane toad, white-winged agouti, opossum, prey, red deer, tatupeba, etc.

Pampas: Located in the extreme south of Brazil, in Rio Grande do Sul, this biome is heavily influenced by the subtropical climate and the formation of the relief, which is mainly made up of plains. Due to the cold and dry climate, the vegetation is unable to develop, being mainly made up of grasses, such as goat-beard grass, fat-grass, mimoso grass, etc. Examples of animals that live in this biome are deer, herons, otters, capybaras and others.

Pantanal: this is the largest floodplain in the country and is located in the states of Mato Grosso and Mato Grosso do Sul. This biome is highly influenced by the river regimes present in these places, because, during the rainy season (October to April), the water of the wetland floods a large part of the region's plain. When the rainy season ends, the rivers decrease their water volume and return to their beds. For this reason, vegetation and animals need to adapt to this movement of water. All these factors make the vegetation of the Pantanal very diversified, with hygrophilic specimens (adapted to humidity), plants typical of the Cerrado and the Amazon and, in the drier areas, xerophilous species. The fauna consists of several species of birds, fish, mammals, reptiles etc.

Preservation of biodiversity in Brazil

The biggest challenge is to reconcile the preservation of different biomes and natural resources with social and economic development. The Director of Conservation and Biodiversity of the Ministry of the Environment (MMA), Bráulio Dias, warns that development without preservation will not give a satisfactory result. The development model, followed by the world in recent decades, created an incalculable environmental and social liability, and, if maintained, creates barriers to national products and is inefficient from the point of view of using natural and environmental resources. The worst risk of this type of development is that it can make economic activity unfeasible. Without understanding that it wears out soils, water resources, generates changes in the climate, a new source of cost is being created to adapt to this depletion of the environment, explains Dias.

According to the MMA director, awareness of preservation has improved in Brazil, but there are more advanced countries in this regard. For him, most of the private initiative and the Brazilian government have not yet incorporated this vision. This can be seen in the conflict between environmentalists and ruralists, fought in the National Congress, about changes in the Brazilian Forest Code, a set of laws that regulate the use of forest areas in the country.

According to the MMA director, awareness of preservation has improved in Brazil, but there are more advanced countries in this regard. For him, most of the private initiative and the Brazilian government have not yet incorporated this vision. This can be seen in the conflict between environmentalists and ruralists, fought in the National Congress, about changes in the Brazilian Forest Code, a set of laws that regulate the use of forest areas in the country.

Bráulio Dias, one of the authors of the study carried out by the MMA, notes that this dilemma between development and protection of the environment has made it increasingly difficult to create new UCs in the country, especially outside the Amazon. Dias explains that the problems sometimes begin with the lack of interest from state governments. He cites the example of the construction of hydroelectric and mining plants, which could face difficulties after the conservation areas were created. "The environment should be seen as an ally. Hydroelectric plants account for 15% of the Brazilian energy matrix and, if we continue to deforest at the edges and headwaters of rivers, there are undesirable consequences, such as the reduction of water infiltration in the soil, erosion, silting up rivers and floods. That is, without preserving, the useful life of hydroelectric plants is reduced and investment is compromised ",

Recognizing the importance of these other biomes, as important areas to protect and not just the Amazon and the remnants of the Atlantic Forest is fundamental. In this case, the caatinga is still so important because it is the only biome that is legitimately Brazilian, he argues. In these places it is important to prioritize the preservation of biodiversity and the recovery and intensification of the use of degraded areas. In this sense, the suggestion would be a strong action to curb the expansion of the agricultural frontier in areas considered ecologically fragile, such as the cerrado and the Amazon.

RESEARCH ON BIODIVERSITY IN BRAZIL

Discovery and characterization of biodiversity - systematics and taxonomy
The traditional system (alpha taxonomy) was based almost exclusively on morphological characters. In the last decades, it started to use more and more tools in the molecular area for phylogenetic studies. In addition, for groups of difficult resolution through morphological characters, it started using DNA Barcoding techniques. In the case of microorganisms, metagenomics has been increasingly used. Taxonomy is a traditional and well-developed area of research in Brazil. In the CNPq research group directory, 255 groups are found working in the area of systematics / taxonomy (148 groups in the plant, 69 in the animal and 38 in the microorganisms), of which 173 use molecular tools in phylogenetic studies, 28 use metagenomics and at least 3 use DNA Barcoding techniques. The main research groups in this area are generally associated with institutions that have large biological collections. These constitute an inexhaustible source of essential information that should, in the future, provide important discoveries that are still out of the technological reach of this generation. They represent genetic banks in which tissue aliquots can be stored, which are essential for studies of molecular biology and biotechnology. They also represent a cultural heritage, testimony to the rich history of the discovery

and expansion of Brazilian society in its national territory, with representatives of the now extinct flora and fauna, which once were part of ecosystems that were irreversibly altered by anthropic action. In this sense, the collections constitute an essential database for studies of characterization and environmental impact.

Understanding the functioning of ecosystems and environmental services Well developed for lentic environments, especially as a result of the research effort initiated in 1970 by the group of Dr. José Galizia Tundisi at the Broa dam (São Carlos), the understanding of the functioning of terrestrial ecosystems is still in the consolidation phase in Brazil. In the CNPq directory there are only 25 registered groups, 13 in terrestrial environments, 10 in freshwater environments and 3 in marine environments. However, very few have worked on the relationship between species richness, ecosystem functioning and environmental services.

The chemistry of natural products in Brazil

Brazil is in a privileged position for the science of natural products, due to being one of the mega-biodiverse countries and having constituted a large body of researchers of chemistry of natural products.[9] In this sense, it is necessary to contextualize the advances in product chemistry in Brazil over the first two decades of the 21st century.Several articles address the history and critical analysis of the chemistry of natural products in Brazil.[6,7,9-14] In a 2001 article, Montanari and Bolzani[15] indicated that it would be desirable to organize chemical and biological information on species from Brazilian biodiversity, to use techniques of dereplication in combination with medicinal chemistry approaches, in addition to the need to improve techniques of herbal analysis, in order to meet the market needs of these products and the need to establish national cooperative networks to strengthen technological innovation initiatives between the academic and the industry.

Brazil, with the greatness of its coastline, its flora and, being the holder of the largest humid tropical and equatorial forest on the planet, it cannot give up its vocation for natural products. Chemistry of Natural Products (QPN) is, within Brazilian Chemistry, the oldest area and the one that, perhaps even today, congregates the largest number of researchers.

The first Portuguese doctors who came to Brazil, due to the shortage, in the colony, of medicines used in Europe, were very early forced to realize the importance of indigenous medicines. Travelers were always supplied with these medicines beforetouring little-known regions.

In addition to the natural remedies used in medical therapy, the dye extracted from the paubrasil tree, the colony's main export product for more thantwo centuries, and one of the reasons for the colonization of Brazil by Portuguese.

The scientific evolution of QPN in Brazil had its historical landmark in classical phytochemistry (isolation and structural determination), implemented and structured by the aforementioned researchers of excellence, from where the consolidated and active groups in the area arose all over the country. This historical panorama can be accompanied by the development of QPN research currently carried out in our parents. It is important to note that several of these research groups have been changing the focus of traditional phytochemistry works involving biological activity, chemical ecology and biosynthesis of plant micromolecules, microorganisms, marine organisms, among others, as well as new analytical methodologies for working with natural products, as a way of acting on the frontier of knowledge, as is the case in countries industrialized.

Medicinal plants

Medicinal plants is a recurring theme on the agenda of Brazilian science. Unanimity among Brazilian chemists and pharmacologists from recognized international expression, studies with medicinal plants have not yet received in Brazil, the attention that the topic deserves funding agencies, although there is already a critical mass of qualified researchers in the fields of chemistry and pharmacology. Up un tilthe present moment, there has not been a coordinated process at allthe actors (industry, pharmacologists, phytochemicals, synthetic chemists, pharmacists, doctors, etc.) aiming at the development of drugs from plants.

The use of medicinal plants is a common practice in the country, which has been transmitted generation to generation (FNP Consultoria e Comércio, 1999) and is carried out through extractivism1 (WWF, 1998). It has its origin in the culture of the diverse indigenous groups that inhabited the country (Simões et al. 1998), mixed, still, with the traditions of use of Europeans and Africans who arrived later and constitutes the current local pharmacopoeia, arousing great interests national and international for the therapeutic and economic potential it represents.

The use and commercialization of medicinal plants has been stimulated, in part, by the growing industry demand for new natural sources of medicines and, on the other hand, due to the side effects caused by synthetic drugs that stimulate the use medicines of plant origin or, in many cases, because they represent the only source of medicines, especially in the most isolated and distant places.

Many plants frequently used by local populations have not yet been studied or active ingredients have not yet been identified to validate them as medicines or to to use them economically (Berg., 1993). Still, many plants are used and currently commercialized and Brazil, one of the countries with the greatest biodiversity in the world, reveals how an important and potential provider of a resource as valuable as plants medicinal. Examples of valuable plants in Brazil include the indigenous Curare or Foxglove (Digitalis purpurea) used in the preparation of tea against dropsy, caused by heart failure before Digitalin's action on the heart muscle was discovered, Bark (Drimys brasiliensis) with stomatal properties, Quina (Cinchona calisaya) used to cure malaria, Ipecacuanha (Cephaelis ipecacuanha) used to treat diarrhea, amoebic dysentery, chronic catarrh, hemorrhage and asthma, and Sapucainha, (Carpotroche brasiliensis) with scientifically proven anti-inflammatory effects and whose oil extracted from the seed is used to treat leprosy (Carrara, 1995).

Popular medical practices in Brazil are dispersed in a multitude of practitioners and users and can be located in any region of the country, surviving the constant harassment of knowledge official doctor, without acquiring the uniformity of scientific medicine since, due to its same origin, uniformity lies in the difficulty of transmitting knowledge, isolation between users and practitioners and ritual prohibitions and the influence of regionalism, creating bounded subsystems (Carrara, 1995).

Many examples of medicinal plants from Brazilian biota could be cited, however, most medicinal plants sold in Brazil are introduced. So the plants endemic medicinal plants are still little known and constitute on a fascinating subject of academic research and development.

Natural products from marine organisms

The oceans cover 70% of the Earth's surface and are inhabitedabout 200,000 species of marine plants and invertebrates andmillions of microorganisms.

In addition to plants, sponges, octocorals, Sea squirts and bryozoans are sessile organisms when adults and develop in adverse conditions when compared to terrestrial

The evolution and survival of these species has resulted in organisms that produce unique substances with different ecological functions.

Research on marine natural products in Brazilbeginning in the 60's at the Natural Products Research Centerat the Faculty of Pharmacy, UFRJ36. However, there are still fewthe information, documented in scientific articles, about isolated substances and the biological activity of natural products from marine organisms collected along the 7500 km of Brazilian coast. The little information available on the chemistry of these organisms, many of which are endemic species, indicate great research potential for the area in Brazil. The history of the study of marine natural products demonstratesthat, around 1960, there was a large investment on the part of thepharmaceutical industries in the search for bioactive substances fromof the organisms that live in the oceans. In addition to the popularization ofscuba diving and the development of equipment forgreater depths, the discovery of large amounts of prostaglandins in an octocoral, aroused interest inresearch on natural marine products.

Since then, several laboratories, mainly in the United StatesUnited States, Australia and Italy, have been dedicated to the chemical study ofsponges, corals, mollusks, algae, microorganisms and other marine organisms. Many identified substances have unique chemical structures that are unprecedented in terrestrial natural sources; fact that has motivated, in addition to the development of newmethods of isolation and organic synthesis, research on the origin and biosynthesis of isolated substances, their ecological importance and pharmacological activities.

The study of chemical substances produced by speciesmarine life is a fundamental tool for understanding the evolution and maintenance of marine communities in differentoceans.

Some of the activities most commonly attributed to natural marine productsthey are those of mediation in reproduction, defense against potential predators, pathogens, bio-encrustation or substrate competitors.only in the last decade did these ecological functions begin to be proven experimentally, including by Brazilian researchers, revealing important roles in the structuring ofecosystems and giving rise to other hypotheses involving the rolesecondary metabolites affecting biodiversity.

Over the past 50 years, nearly 10,000 natural marine products have been discovered, many with pharmacological activity, including acetogenins, polyketides, terpenes, alkaloids, peptides and metabolites of mixed biosynthetic origin.

Studies on sulfated polysaccharides with activities anticoagulant and antithrombotic, isolated from algae, ascidea, cucumbers and sea urchins, are the most significant in the area of natural marine products in Brazil. As an example of the results of these researches, chondroitin fucosylated sulphate, isolated from tame cucumber Ludwigothurea grisea, presents itself as a promising alternative to heparin, due to its anticoagulant and antithrombotic activities49. The difficulty in cultivating marine macroorganisms or synthesis of complex molecules on a large scale have hindered the development of drugs of marine origin. On the other hand, several substances with pharmacological activity are suspected of being produced by associated microorganisms, capable of large-scale culture. This explains, at least in part, the interest in the study of marine microorganisms isolated from sediments, seawater, macroalgae, fish and invertebrates.

The marine environment emerges as an important natural source due to its fantastic organic diversity, which remains practically unexplored. Currently, chemical and biological approaches of marine organisms represent a wide and promising area of research, given the constant discovery of several metabolites with varied medicinal properties, in addition to a practically unlimited metabolic arsenal. Red algae, especially the Rhodomelaceae family, are excellent producers of halogenated metabolites to which important biological activities are attributed. Marine and / or endophytic microorganisms are identified as the most promising targets for the discovery of new drugs.

Among the marine species responsible for providing chemically intriguing natural products, marine cyanobacteria (blue-green algae) are possibly the most successful in terms of producing new chemotypes with potent biological activity . Among anticancer candidates, apratoxin D stood out for exhibiting an IC50 of 2.6 nM (minimum concentration capable of inhibiting the proliferation of 50% of tumor cells) when evaluated in human lung cancer cell lines H-460. Jamaicamides A-C, isolated from Lyngbya majuscula, showed cytotoxicity in human lung cancer cell line H-460 and neuroblastoma in mice (Neuro-2a) . Recently, the new lipids serinolamide A and propenediester, were isolated from the species Lyngbya majuscula (collected in Papua New Guinea) and Oscillatoria sp. (found in Panama), with serinolamide A being the most current representative among cannabinomimetics derived from marine origin.

Among the universe of microorganisms, it is also important to highlight the chemistry of marine fungi, which have been found associated with algae, plants, invertebrates, molluscs, and even present in sediments from coastal areas and mangrove regions (11). As examples, we can mention the drimanic sesquiterpenoids, isolated from the fungus Aspergillus ustus (present in the marine sponge Suberites domuncula), cytotoxic against several tumor cell lines and the AC sporotrins, exhibiting strong inhibition of acetylcholinesterase, isolated from Sporothrix sp .. Another very fascinating molecule found in Phoma sp was epoxyphomaline A, which in addition to being cytotoxic to 12 of the 36 human tumor cell lines evaluated, demonstrated a specific mechanism of action not correlated with those exhibited by patterns of anticancer agents.

Despite being a very attractive area, there are still some limiting factors regarding the research and bioprospecting of marine organisms. Often the process becomes quite long due to the methodologies used for the isolation, structural clarification, characterization and management of the results obtained. These factors are aggravated by the structural complexity, the exceptionally limited amount available for studies and the difficult conditions for collecting and handling this type of material. Additional complications arise regarding the identification (chemotaxonomic aspects) of the organisms, and also due to the uncertainty regarding the origin of an active compound, which may have been synthesized due to the symbiosis phenomena.

Natural products of microorganisms

Pharmaceutical products together with agrochemicals are today considered the two supporting pillars of modern civilization. The phrase mens sana in corpore sano is certainly the product ideal of an interdisciplinary scientific project, the final result of which is the quality of life of the human species. In this context, products isolated from microorganisms, in general, have a unprecedented importance not only as medicines (example antibiotics), but mainly as agrochemicals less harmful to human health.

One of the most important properties of microorganisms, fungi in particular, is associated with their metabolic capacity to produce a great diversity of bioactive micromolecules. Nevertheless, fungi are also responsible for the production of highly toxic substances to mammals, known as mycotoxins, some considered to be potent carcinogens.

In tropical environments, fungi are a serious problem in the food storage process, mainlyof cereals, due to the production of toxic mycotoxins that contaminate food and become a serious public health problem. On the other hand, important drugs of clinical use in severalpathologies were obtained from fungi53. This dichotomy of functions cancome from the great chemical diversity that fungi produce.

With regard to occurrence and biodiversity, fungi are the second largest group of species on earth, losing only for insects. Estimates suggest that there are approximately 1.5 million different species of fungi, less than 5% were described

Thus, in recent years there has been a research priorityon the chemistry of fungi, fearing the loss of biodiversity.Many endangered species of higher plants and insects are associated with the specific flora of the fungi and, if these losses occur, they may result in the disappearance of the fungal species.The search for bioactive principles of microorganisms is one of theareas in which most is invested in developed countries, mainly in bioprospecting research carried out by industriespharmaceutical companie.The use of products obtained from fungi in the biological control of agriculture has grown markedly. In this process, the use of certain fungi as mycoerbicides is highlighted, mycoinsecticides or mycoparasites.Due to the production of large quantities and diversity ofsecondary metabolites, competitive fungi are considereda valuable source of products with pharmacological activity.

When a microbial metabolite is consideredcandidate for a new drug, its production can be carried outon a larger scale, using the manipulation of parametersof the fermentative process for the expression of metabolites, aiming to improve your income. The raisein the production of these compounds can also be achieved, submitting the lineage of the microorganism to programsimprovement. You can isolate mutants more easilycultivable, which generate additional products ormodified with higher therapeutic index.

Microbial fermentation as a means of production bioactive substances has several advantages, such as possibility of reproduction and safe productivity, since the microorganism grown in fermentation tanks becomes a potentially inexhaustible source. The increase in production relatively easy, with changes in working conditions culture can be explored in order to optimize the various biosynthetic pathways that can lead to the production of compounds even more effective. In addition, microorganisms respond favorable to routine culture techniques, while tissue cultures or plant growth require specialized techniques or months of growth before collect.

Microbial biotransformation is an alternative method for obtaining bioactive compounds30 using systems enzymatic biological agents to produce chemical changes in compounds that are not their natural substrates31. An molecule can be modified by transforming functional groups, resulting in the formation of new products and useful that are not easily obtained by chemical methods. The use of microorganisms in this process is due to the its unlimited ability to adapt to new environments and metabolization of various substrates.

The biotransformation process provides advantages over chemical synthesis, as it can be performed at temperature environment and without the need for high pressure and conditions extreme conditions, thus reducing unwanted by-products, energy needs and costs.

Endophytic fungi are frequent sources of productsthat can act as antibiotics, inhibiting orkilling a variety of disease-causing agentsharmful substances such as bacteria, fungi, viruses and protozoaaffect humans and animals. In addition to antibiotics, severalhigh value-added drugs can be produced fromof endophytic microorganisms, extracted from a smallportion of plant tissue, thus maintaining the productionof compounds vital to people affected by countlessdiseases. The use of endophyte metabolites in the industrypharmaceutical industry became evident in the 20th century, discovery of the production of the diterpenoid taxol by the fungusendophytic Taxomyces andreanea. This antitumor compoundwith cytotoxic activity and high international value wasoriginally found and extracted from plant species in thegenus Taxus. With this discovery, it became possible toproduction of this important drug more efficientlyand less expensive, minimizing the threat of extinction ofsome plant species collected for the extraction of thiscompound and its precursors. Currently, this drug thatinterferes with the multiplication of cancer cells, reducingor interrupting its growth and dissemination, it is alreadyused for the treatment of breast, lung andovary.

Chemistry of natural products and their relationship with plant protection

Plants can have constitutive mechanical defenses for resist the attack of herbivores, such as: resins, increased concentration of lignin, accumulation of silica and wax in the epidermis, which alter the texture of the fabrics, thus reducing their palatability and digestibility for the herbivore. Surface barriers that restrict the movement of the herbivore, such as thorns and trichomes, also may be present (DUDAREVA et al., 2006).

In addition to the constitutive barriers, plants have the capacityto change its physiology, its development and also itschemical composition when subjected to attack, a mechanism thatknown as induced defense. The induced defense is a setdynamic phenotypic responses that allow plants todefend only when necessary, as this is a mechanismwith high energy consumption and its continuous activation cancompromise the allocation of the necessary sources for growth andreproduction (AGRAWAL, 1999).

The importance of thistheme can be assessed by the increasing destruction of cultivarsof cereals by the attack of pests, which causes the loss of 1/3 of theworld cereal production. This problem has been exacerbated byincreasingly frequent practice of monocultures and the resistance ofpredators to synthetic insecticides. The search for natural insecticides gained enormous momentum afterthe discovery of the undesirable effects on the ecosystems of synthetic insecticides, which has DDT as its main villain. Plants have their own defenses that protect them from other plants, phytophagous insects and predatory herbivores in general. These defenses are chemical in nature and usuallyinvolve substances of secondary metabolism, which maybe called phytotoxins or allelochemicals. This phenomenon is known as allelopathy. The first plant-plant interaction was described by the Roman naturalist Plínio, who observed that under the canopyof walnuts do not grow other vegetables.

The most recent definition of allelopathy, one that is accepted by the International Allelopathy Society, is very comprehensive, as can be seen below: "It is science that studies any process mainly involving secondary metabolites produced by plants, algae, bacteria and fungi that influence the growth of biological systems with positive and negative effects". Such a broad definition for allelopathy makes it difficult to even state what are allelochemicals.

Vegetables develop defense mechanisms against pathogens (viruses, bacteria, fungi, insects, etc.) producing toxins againstinvading agent and acquiring resistance to infection. Since the decade60, these mechanisms are known, but only in the last few yearsyears, plant pathologists have been dedicated to the study of the basesmolecular and genetic factors of this phenomenon.

Recently the involvement of salicylic acid was discovered(AS) and its acetylated derivative (AAS) in the defense reactions againstpathogens.

Salicylic acid is accumulated in the plant tissue afterinfection, causing an immune response, called resistancesystemic acquired (SAR).

Another substance with a very simple and important chemical structure in the plant defense system and which has been the subject of studies by plant biochemists is methyl jasmonate (32). This substance induces the production of enzyme inhibitors that break down proteins (proteinases), antifungal proteins and enzymes involved in the biosynthesis of secondary metabolites from defense (Scheme 1). In this way, in addition to the functions already well known in the development and regulation of plant growth, jasmonates play an important role as phytoormones or transducers in plant defense signaling.

Another example that shows the beauty of the plant-plant adaptive process and how the Natural Products chemists with their research can bring great benefits to humanity is related to the weed Striga asiatica, one of the most devastating pests of cereal crops in the world. This parasitic plantdecreases the food reserves of almost half a million people in theAsia and Africa. Striga asiatica has a true chemical radar and only germinates and can remain latent in the soil for a long time, when the host plant releases estrigol. At the time ofrelease, the parasite clings to the host's roots and feeds on it.

The secondary metabolisms of plants are basically the organic compounds produced for the plant's own protection. When the plant wants to protect itself from a microorganism, or even an insect, it itself produces a substance that can kill, inhibit or even repel that insect or microorganism.

Cunila angustifolia that its oil has already been tested for insecticidal effects and has shown good results. A research on the application of Cunila angustifolia oil was carried out in an aviary bed, with a reduction of 60% of the insects present there. The same oil was also tested with woodworm that occurs in the crops of corn and beans. In laboratory tests, this oil inhibited 100% of the bean weevil, with an insecticidal effect, killing all insects. Therefore, this is a promising new natural product for use both in the storage of corn and beans to replace products containing pyrethroids or organophosphates currently used in conventional food production.

Euterpe edulis, which is better known as juçara, or false açaí. Currently used basically for the extraction of the heart of palm, which is very valuable in the market, but which leads the plant to death after removing the nucleus from its stem. The use of alternative parts of the plant, such as its fruits, can help to reduce irregular exploitation that includes it among Atlantic forest plants with risk of extinction. Its presence in the forest is also important for the maintenance of a series of birds that feed on its fruits. The fruits of this plant are very rich in anthocyanins, which is a group of compounds already proven to be very efficient for the treatment of a series of diseases, with excellent results also against stress and other oxidative actions. Research is also underway on its use as a natural dye, which can be useful both for the food industry and for use as a paint for children, as it does not present toxicity like some paints available on the market.
The natural compounds extracted from juçara palm fruits, with their high antioxidant potential, are also being tested for use as films, for coating and external protection of food.

The role that plant VOCs play in direct defenseof the plant is to promote defense against abiotic and biotic stresses. This function is probably a reflection of the ancestral function thatthese compounds played in heterotrophic bacteria andphotosynthetic, before joining the first eukaryotes toform the first order of plants. Some volatiles, such as volatilesGreen Leaf (VFV), for example, are antimicrobials that protectfruits and leaves from attack by some pathogens. In addition, ELVsThey can also act against some microorganisms that areon the surface of the plant (BALDWIN et al., 2010).

Plants have a unique ability to recognize the attack of herbivores and can reconfigure their entire response to produce a lot of defense compounds, such as metabolites toxic side effects that act as a direct defense mechanism. Among these metabolites that are present in plant tissues include allelochemicals with antinutritional, toxic or repellents to the herbivore, and as an example one can list the glycosides cyanogens, digestive enzyme inhibitors, lectins, glycosinolates, alkaloids and terpenoids. Indirect defense consists of the production volatile organic compounds (VOCs) that are released by plants when they suffer some damage caused by the herbivore and can attract predators or parasitoids that will protect the plant or repel oviposition or herbivorous insect (VOELCKEL and BALDWIN 2004).

Other volatile compounds that are described with properties ofdirect defense of the plant are volatile toxins. These, almost alwaysare produced and stored in the form of non-toxic conjugates, and when there is a rupture of the tissue after some damage, they will be releasedin large quantities. As examples of these toxins can becited are cyanides and isothiocyanates and their conjugated formsin the form of cyanogenic glycosides and glycosinolates respectively, that are present in several plant species.

One of the first danger signs for plants is the laying of eggson its surface, as it is very likely that there will be an emergence oflarvae and consequently the attempted damage. Plants realizethat the eggs are deposited and respond by activating a series ofmechanisms of both direct and indirect defenses. How much is depositionof the egg there may be the induction of the formation of neoplasm that willthe egg from the plant's surface, making it easilydetached from the site and fall off the plant (DOSS et al, 2000).

Plants also have the ability to necrosis the tissue where the eggs are deposited, causing the eggs fall from the plant and thus the development of the bug

The emission of VIPHs at the plant is induced by elicitors (or elicitors) that are present in the insect's saliva is generally species-specific and therefore is one of the main determinants ofspecificity of the response (DE MORAES et al., 1999) and knownfor promoting the activation, properly speaking, of the mechanisms ofdefense. Different elicitors have been described in the saliva of differentinsect species, including: conjugated lipo-amino acids such asN-17-hydroxylinolenoyl-l-Glutamine (Volicitin), sulfated fatty acids(caeliferins), some enzymes and protein peptide fragmentsof the plant (Inceptinas) (ALBORN et al., 2007; SCHMELTZ et al., 2007).

The composition of the different volatile mixtures will also dependthe type of damage the plant is suffering, that is, if the damage wascreated by oviposition or through the attack of chewing herbivoresor sap suckers

bioprospecting in Brazil

One of the ways to extract economic value from biodiversity is to bioprospecting. Here it is defined as the systematic search for organisms, genes, enzymes, compounds, processes and parts from living beings in general, which canhave an economic potential and eventually lead to product development.It is relevant to a wide range of sectors and activities, including biotechnology,agriculture, nutrition, pharmaceutical and cosmetics industry, bioremediation,biomonitoring, health, fuel production through biomass, among others others. The targets of bioprospecting are collectively called genetic resources. Together, they form the national genetic heritage. The process that goes from the identification of a biological component for usepotential for a commercial product is complex, involves many steps and is typicallylong, costly and risky, usually requiring a lot of technology and highly staffedqualified for research.

The main political challenge, therefore,is the effective integration between biodiversity, biotechnology, infrastructure,industrial, innovation and intellectual property. Winning it is fundamental for thepublic and private capital come together to make the biotechnology industrybreak away from technological dependence and use available natural and human capital,to generate income and reduce inequalities in a sustainable manner.

Bioprospecting, now known as biodiscovery, 20 aims at investigating bioactive molecules with the purpose of describing new drug models. Since the 1970s, initiatives of this nature in the Brazilian academic environment have been discussed.9 However, the only really well-established biodiscovery program initiative is BIOprospecTA, part of the BIOTA-FAPESP program.9 BIOprospecTA allowed a certain organization of research groups in chemistry of natural products in the state of Sao Paulo, and represents a real milestone in the development of the chemistry of natural products in Brazil, as it allowed PN research in Sao Paulo to be strengthened, gain international visibility, establish collaborations with numerous research groups, offer opportunity for students to do internships abroad, hold numerous workshops, symposia and mini-courses, as well as establish partnerships with the private sector for the development of bioproducts. Although significant advances are desired, these should arise with the maturation of BIOprospecTA, in order to further strengthen research in this area in the state of Sao Paulo. BIOprospecTA is one of the few, if not the only, initiative formally established and coordinated by chemists in Brazil, which further reinforces its innovative character. In the last 15 years, there has also been an increase in funding for Brazilian science. Several natural product chemistry groups in the North, Northeast, Midwest and South have organized excellent laboratory structures for research in the chemistry of natural products. Incentive programs for research on biodiversity and bio-discovery, such as CNPq's SISBIOTA-Brazil, strengthened research on natural products in Brazil, contributing to the establishment of new groups and the strengthening of others that are still in their infancy. In addition, two National Institutes of Science and Technology are focused on research in natural products, the National Institute of Science and Technology for Biorrational Control of Pest and Phytopathogenic Insects and the National Institute of Science and Technology in Biodiversity and Natural Products. These INCTs bring together numerous research groups on natural products, with significant funding, and bring perspectives for the discovery of new molecules from different natural sources. Other INCTs and a CEPID funded by FAPESP, CIBFar (Center for Research and Innovation in Biodiversity and Pharmaceuticals), develop projects in the field of medicinal chemistry and discovery of drugs, antioxidants and other bioactive molecules. These initiatives contribute significantly to strengthening the chemistry of natural products in Brazil.

Products naturals chemistry for society

The financing of scientific research in Brazil is going through a complex moment, in which priorities are reviewed and much question is asked about the quality of Brazilian science. The research carried out in Brazil is found, in 69% of its extension, in Brazilian universities.36 In addition, the share of academic research that is financed with resources from the federal and state governments is much greater than from the private initiative.36 considering that public funding for research comes from tax collection, the direct relationship between funding for scientific research in Brazil and any benefits that society has the right to claim is evident. In addition to being a complex topic that arouses the most diverse opinions, it is not possible to escape one point: Brazilian scientists owe a debt to society. In this sense, researchers of chemistry of natural products must be attentive to the demands of society with regard to the quality of basic research developed, the solution of problems that arise, as well as the offer of products that can benefit society.

Society, the national pharmaceutical industry and academic researchers still lack a much more significant integration, which allows bringing innovations of different natures - processes, products, technologies, education and professional training - to the context of scientific development Brazilian.

Products naturals chemists will be important players on the stage of the future. The dominance of plants will be possible when the magical world of plants enzymes is better understood and metabolic pathways fully known. Expressing or silencing genes will be routine activity in molecular biology, when biologists and chemists would seemingly peer into the bottom of living plant cells, their nuclei, chromosomes and genes92. The journey in this direction has already begun and the participation of Natural Products chemists will be increasingly important for this path to be taken faster. O domain of analytical techniques and separation preparations, both secondary metabolites such as plant macromolecules such as techniques for identifying these same molecules, is part of the routine of Natural Products chemistry in industrialized countries. This does not mean that the Brazilian Natural Products chemistry community abandons its drug bioprospecting projects new to engage in code-breaking mega-projects genetic factors or stop looking in plants for the reason of their resistance pests. But it is also imperative that new groups of phytochemicals don't just keep doing what they've learned with their masters to stop being scientific clones

Chemistry of natural products in the genomic era

Man has always sought to understand the nature of life and the chances of achieving it and, at least at the molecular level, has never been closer. New information on genes and their functions, in part, are responsible for a technological revolution aimed at life sciences. It is now up to the Brazilian chemist to stop and reflect on what we really need to know in order to move from compiling data to understanding it. In this text, we will comment and show that the moment is propitious for start modeling the metabolic profile of plants in our biota, mainly those of economic interest such as drugs and / or agrochemicals.

Genomics, a term that in the last two years has occupied allmeans of communication, is the activity of sequencing genomes and deriving theoretical information from the analysis of the set of expressiongenetics of an individual, using computational tools, while functional genomics defines the status of the transcriptome and theproteome of a cell, tissue or organism under defined conditions; in other words, it determines the function of genes in an organism. The ultimate goal of proteomic research goes beyond simplecataloging of proteins that a cell expresses in its statenormal or sick. The main objective is to elucidate the regulatory networkand signaling the organization and metabolic dynamics by which thecell life is processed.

Plant genomics has not yet reached the status of genomicsof microorganisms and animals, however, with the sequencing of thegenome of Arabidopsis thaliana, the first plant species to have itscompletely deciphered genetic code, a new phase beginsin research on plant molecular biology and mainlyon the knowledge of metabolic regulation genes.

Chemistry of Natural Products in the post-genomic era continues being a fundamental tool for understanding the metabolic mechanisms of each part of the cell and the functions of these metabolites in a given cell unit90, even though we know that none of these substances are involved in the basic metabolic processes of the cell, but that they account for signaling, adaptation, pollination and defense. The most complex and fascinating in the study prospective of plants is that each taxon has its own system, characterized by specific metabolites: alkaloids, terpenoids, phenolic, etc.

Within this premise, about 100,000 substances plants are known and it is estimated that about 4,000 new plants are described each year.

Plant nature will always be an inexhaustible source of useful substances, such as secondary metabolites, but almost alwaysproduced in insufficient quantities for any economic utility. In view of this fact, the cellular mechanisms of regulation andproduction of these substances may be modified so that they are produced in natura, by interference in the responsible genesby the metabolic regulation of a species of interest, or beproduced by programmed cell culture.

In the post-genomic era, with all the advances in biological knowledge, both academic laboratories are rediscovering the value of modulatory compounds to explore biological processes and the pharmaceutical industries are considering biological research based on chemical "probes", in order to find new therapeutic targets and also small molecules that allow their modular activities.

Chemical Genetics can be defined as the systematic use of small molecules to alter the function of proteins to which they bind, exploring recent advances in robotics, combinatorial chemistry, large-scale assays (HTS) and bioinformatics. In summary, Chemical Genetics involves a genetic study using chemical tools. It is a complementary approach to genetics, widely used in biology, that alters the function of proteins through mutations in genes of interest, whether by inactivation (deletion or "knock-out") or activation (oncogenic). Chemical Genetics involves the exposure of cells to a collection of compounds of any origin (natural or synthetic) for the selection of the molecule that induces the phenotypic change of interest, followed by the identification of the protein responsible (macromolecular target) for the phenotypic change. Thus, chemical genetics operates in the direction of the effect for the cause, that is, from the phenotype to the genotype, and the mechanism of action of the compound is determined later. This is the oldest drug discovery paradigm.

The term Chemical Genetics was introduced due to the specificity with which some natural products interact with proteins, leading to effects comparable to knock-outs in genes. Therefore, some natural products are powerful tools for the study of biological processes, and can contribute to the third challenge of Chemical Biology through the understanding of biochemical pathways and the identification of new proteins. Although historically successful, this phenotypic assessment approach has some limitations for natural products, especially when looking for active substances. Only extracts containing the most potent and / or abundant substances will show positive results, while less abundant or moderately active compounds, although potentially interesting, may not be detected. In addition, cytotoxic compounds can mask subtle phenotypic effects of other compounds present in the crude mixture. Finally, synergistic or cooperative effects between compounds in mixtures can be lost after fractionation. To mitigate these limitations, simple pre-fractionation of the extracts can be carried out using adsorbent resins such as Diaion HP-20, XAD, C-18 or even silica gel, which lead to the achievement of less complex fractions. Some laboratories also use parallel preparative HPLC systems to separate a large number of crude extracts, allowing relatively pure samples to enter the initial screens, increasing the possibility of successful identification of new bioactive products.

HISTORY OF THE ROLE OF NATURAL PRODUCTS IN DRUG DISCOVERY

Several Native American civilizations used body and hair painting as a means of communication, before any reports of writing or paintings in caves and sacred sites. Natural dyes, such as bixin genipin and andirobin were used for aesthetic, religious and protective purposes; other valuable products such as balms, gums and essences were also appreciated and useful as environment repellents and odorized,

The history of Brazil is closely linked to the trade in natural products (spices), which determined the various disputes over possession of the new land and, finally, Portuguese colonization. Redwood (Cesalpinia echinata) produced a red dye, widely used for dyeing clothes and as writing ink, which was known and used in the East Indies since the Middle Ages. Brazilina was extracted from the wood of brazilwood , a catecholic derivative that easily oxidized to brazilein , a phenoldienonic identified as dye.Until the end of the 19th century, only natural dyes were available, making these products valuable and of enormous interest to the colonizers. In this sense, in addition to the brazilwood, which was extracted predatory from our territory until almost its extinction, many other products aroused interest on the part of Europeans who arrived in the newly discovered land. Morina , obtained from Chlorophora tinctoria, was another natural dye exported to Europe, remaining prominent in the trade until the development of aniline chemistry in Germany and, until today, it is used as a sugar indicator in chromatography thin layer .The deep knowledge of the chemical arsenal of nature, by primitive peoples and indigenous people can be considered a fundamental factor for the discovery of toxic and medicated substances over time. Coexistence and learning with the most different ethnic groups brought valuable contributions to the development of research on natural products, knowledge of the intimate relationship between the chemical structure of a given compound and its biological properties and the animal / insect-plant interrelationship. In this sense, nature provided many molecular models that underpinned studies of structure-activity relationship (SAR) and inspired the development of classic organic synthesis. There are several examples that could illustrate this extensive and fascinating subject. The healers were drugs obtained from several species of American and African Strychnos and Chondodendron, used by the Indians to produce poisoned arrows for hunting and fishing. The first curare plant identified was collected in Suriname and described, in 1783 by Schreber, as American Toxicaria, having been later classified as Strychnos guianensis.

Studies resulting from the understanding of the relationship between the chemical properties of a dye and its chemical structure led Claude Bernard (1856) to deduce that curare acted as a neuromuscular blocker6. Due to the structural complexity of curare alkaloids, only 12 years after the characterization of tubocurarine, the first quaternary ammonium salts appeared as ganglionic blockers6. Curare was also responsible for the initiation of studies on the relationship between chemical structure and biological activity (SAR), being, in this area, the first published work on SAR in Pharmaceutical Chemistry, dated 18696.Another striking example of natural products that had a great impact on humanity, and that in some way changed the behavior of modern man, was the discovery of hallucinogenic substances. Ancient people used snuff and hallucinogenic drinks to a great extent in their religious and magical practices. In ancient Greece, plant extracts were used in executions, as in the case of Socrates, who died after ingesting a hemlock drink, which contained the conine.

Opium, prepared from the bulbs of Papaver somniferum, has been known for centuries for its soporific and analgesic properties. This plant has been used since the time of the Sumerians (4000 BC), with reports in Greek mythology attributing to the opium poppy the symbolism of Morpheus, the god of sleep7. In 1803, Derosne described the "opium salt", beginning the first studies on the chemical constitution of opium; in 1804, in France, Armand Séquin isolated its major constituent, morphine (, and Sertürner published his works on the principium somniferum, having been the pioneers in the search for the use of natural substances in pure form. Opium also produces other alkaloids with interesting properties such as codeine, antitussive, tebain, morphine antagonist, narcotine, antitussive and spasmolytic and papaverine.

During the Spanish colonization of Peru, in 1630, the Jesuits became aware of the use by the Indians of the dry barks of Cinchona species to treat some types of fever9. In 1820, Pelletier and Caventou isolated quinine, which for almost three hundred years was the only effective active ingredient against malaria.

In the beginning, chemists studied plants consecrated by popular use, generally incorporated into the pharmacopoeias of the time, limited to the isolation and structural determination of active substances10. Given the importance of plants for medicine at the time, Chemistry and Medicine started to have a close relationship, which allowed a rapid development of their specific fields10. In this way, many active substances were known and introduced into therapy, and remain today as medicines. Some examples have already been described, such as the Cinchona and Papaver alkaloids. The advance in importance of science and technology brought about profound social and commercial changes, culminating in the Industrial Revolution, which occurred in the 19th century. Perhaps the most important milestone for the development of drugs from natural plant products was the discovery of salicylates obtained from Salix alba. This fascinating story begins in 1757, when the Reverend Edward Stone tasted the bitter taste of the willow bark (S. alba) and associated it with the flavor of Cinchona extracts. This fact piqued his curiosity and imagination, prompting him to report to the Royal Society, six years later, the results of his clinical observations showing the analgesic and antipyretic properties of that plant's extract

Natural products and their commercial use

Since the beginning, man has used natural products in his daily. He used them to simulate nature, to dress up and beautify himself to parties or battles through pigments of natural origin. Extracted poison from amphibians in hunting aid or made use of hallucinogenic plants or fungi in rituals religious and healing. From the moment that man began to understand and use natural products, in its daily routine, the history of the development of science and technology in several areas, including food, textile and pharmaceutical industries. It is simple to think that men make use of natural resources as food, in therapies or for beautification, either through plants, minerals or even animals. However, the use of substances isolated from nature began to develop as new technologies have been developed to separate a single product, in pure form, from a set of a certain natural source. From a therapeutic point of view, for example, about 80% of the population has contact with some traditional therapy, which invariably involves natural sources of treatment. In this case, phytotherapy is one of the most important used in the treatment of diseases.

Since the last century, natural products or their products have been derivatives launched on the drug market.
Of the medicines of natural origin we can mention the quinine, for thetreatment of malaria, and morphine, as an analgesic, among alkaloidsfrom plants.

Antibiotics, such as penicillins, are obtained from Penicillium sp., In industrial fermentation processes that use the microorganism itself to develop them in large quantities. Its antibiotic activity was extremely useful for prolonging human life on Earth. Among the natural penicillins most common, penicillin G (benzylpenicillin), with extensive action on bacteria, was very successful in the pharmaceutical market.
In Brazil, many products and natural sources are researched for thediscovery of new drugs. A historical example in the country is obtainingof lapachol from species of ipe-purple. Lapachol, a naphthoquinone, isvery

effective in treating infections and cancer. This, among othersinformation suggests that natural products are important models forresearch of new therapeutic agents.

However, not only drugs are obtained industrially from nature. Flavors and fragrances are other extremely important products for the cosmetic, food, chemical and pharmaceutical industry.

Substances related to flavor and aroma move billions of peopledollars a year, around 25% of the total food market in the world. Although most flavorings are obtained synthetically, the structureschemicals are identical or were designed from natural products.

Among the main flavorings used in the food industry andcosmetics, the most common ones are vanillin, benzaldehyde, cinnamaldehyde - thethree aldehyde-type compounds - and methyl anthranilate, an ester.

It is possible to think of sources of natural dyes, such as annatto and saffron, which have different pigments in their composition and are used like culinary spices.

Algae are also sources of thickening agents, such as carrageenan, thealginate and agar, widely used in the food and cosmetic industry. The textile use of natural dyes has existed for over four thousand years. Obest known example of a natural dye is indigo, used well beforeChrist. It is extracted from plants of the genus Indigofera, native to tropical countries andsubtropical.

Bioactive natural products

Bioactive natural products have become known from studies on chemical diversity and the role of each metabolite in natural matrices (plants, microorganisms or marine organisms). They can be compared to an orchestra that, with impeccable precision and skill, transform notes of nature into endless chemical possibilities with potential biological interest.

Bioactive molecules come from the purification of natural extracts using techniques of separation, such as chromatography. The phytochemical study of these molecules has been developing due to the pharmacological potential, thus representing an alternative to therapies conventional.

Bioactive natural products are economically important as drugs, fragrances, pigments, food additives and pesticides. The biotechnological tools are important to select, multiply, improve and analyze medicinal plants for production of such products. The utilization of medicinal plant cells for the production of natural or recombinant compounds of commercial interest has gained increasing attention over the past decades. Plant tissue culture systems are possible source of valuable medicinal compounds, fragrances and colorants, which cannot be produced by microbial cells or chemical synthesis. *In vitro* production of bioactive natural products in plant cell suspension culture has been reported from various medicinal plants and bioreactors are the key step towards commercial production. Genetic transformation is a powerful tool for enhancing the productivity of novel products; especially by *Agrobacterium tumefacians*. Combinatorial biosynthesis is another approach in the generation of novel natural products and for the production of rare and expensive natural products. Recent advances in the molecular biology, enzymology and bioreactor technology of plant cell culture suggest that these systems may become a viable source of important secondary metabolites. Genetic fingerprinting could be a powerful tool in the field of medicinal plants to be used for correct germplasm identification. In addition, when linked to emerging tools such as metabolomics and proteomics, providing fingerprints of the plant's metabolites or protein composition, it gives data on phenotypic variation, caused by growth conditions or environmental factors, and also yield data on the genes involved in the biosynthesis. DNA profiling techniques like DNA microarrays serve as suitable high throughput tools for the simultaneous analysis of multiple genes and analysis of gene expression that becomes necessary for providing clues about regulatory mechanisms, biochemical pathways and broader cellular functions. New and powerful tools in functional genomics can be used in combination with metabolomics to elucidate biosynthetic pathways of natural products.

Isolation and Identification of Natural Products

The isolation of the compounds can be done in several ways. The most common involves the use of chromatographic methods. Initially a plant extract is obtained through the use of organic solvents. This extract is then subjected to chromatographic separation of its components until the compounds of interest are obtained in pure form. Once isolated, the compounds have their structures determined through instrumental methods.

Extraction is a relevant step in the Chemistry of Natural Products and the correct choice of method represents success in the isolation of plant metabolites. Therefore, knowing the extraction methods available, as well as assessing how the related variables interfere in the process is of paramount importance for the natural product chemist (FERRO, 2006). Among the extraction methods the most used is maceration, which consists of placing the pre-established plant material in contact with the solvent to be extracted, in a closed container, at room temperature, for 2 to 14 days. It is a procedure that results in a balance of concentration between the plant material and the solvent used, the solvent will meet selectivity, viscosity and polarity, and the plant material will vary in terms of the amount in weight, degree of humidity and particle size (ANVISA , 1996; SIMÕES, 2002). The Melastomataceae family has about 4,200 to 5,000 species belonging to 11 tribes and 185 genera, constituting one of the most important families of the neotropical flora (BAUMGRATZ; CHIAVEGATTO, 2006). Among the genera that occur in the world, the most diverse is the genus Miconia, which also appears as the largest in number of species and has still been little studied from the biological and chemical point of view (CADDAH, 2016). The species Miconia minutiflora, native to Brazil, popularly known as the small leaf inkwell, is found in the North, Northeast, Midwest and Southeast in the biomes, Amazon, Caatinga, Cerrado and Atlantic Forest in the form of a tree. Adsorption chromatography: Uses a solid stationary phaseand a mobile liquid or gaseous phase. The solute is adsorbed onsolid particle surface. The balance between FE and FM justifies theseparation of the different solutes.

Partition chromatography: A liquid EF forms a thin film on the surface of a solid support. The solute is in balance between FE and FM.

Ion Exchange Chromatography: anions like –SO3 or cations as –N (CH3) + 3 are covalently linked to solid FE, usually a resin, in this type of chromatography. The ions of opposite charge of the solute are attracted to the FE by the force electrostatic. FM is a liquid.

Molecular exclusion chromatography: Also called chromatography gel filtration or gel permeation, this technique separates molecules by size, with the largest solutes passing through it with greater speed. Unlike the others types of chromatography, there are no attractive interactions between EF and solute in the real case of molecular exclusion. More precisely, FM liquid or gas passes through the porous gel. The pores are small enough to exclude large molecules from the solute, but not small ones. The flow of large molecules passes without entering the pores. Small molecules take longer to pass through column, because they enter the gel and therefore must go through a higher volume before leaving the speaker.

INSTRUMENTAL METHODS

Instrumental analytical methods consist of measuring the physical properties of the analyte, such as conductivity, electrode potential,absorption or emission of light, mass / charge ratio and fluorescence. In thesemethods involve the use of sophisticated equipment, but alsoit may involve chemical reactions in some stages. They are often less accurate than classical methods, although they are faster. Areused in quantifying and identifying minority constituents.Chemists started exploring other related phenomenawith the properties of analytes in the 1930s, more precisely in the middle.The measurement of the physical properties of the analyte began to be used toquantitative analysis of a variety of organic, inorganic andbiochemicals. A little later, chromatographic techniques appearedthat replaced the distillation, extraction and precipitation of components incomplex mixtures, before qualitative or quantitative determination. Thesenew methods for the separation and determination of chemical species areknown together as Instrumental Methods of Analysis Instrumental analytical techniques have excelled in recent years years for technological advances, both in the assembly of analytical systems more robust and smaller, and in software development operation and data processing that optimize analysis and interpretation of the results obtained, which have led, in association with other market factors, to a reduction in the prices of laboratory equipment.

Electrochemical Techniques

Electrochemical, or electroanalytical techniques, are based on oxidation-reduction processes where the electroactive species in the environment respond to the application of an electrical potential for the monitorching of the electric current, or to obtainof the value of the potential of the analyte in comparisonto the potential of a reference electrode. Of thesetechniques can stand out Amperometry, theVoltammetry and Potentiometry.

Spectroscopic Techniques

Spectroscopic techniques have the answeranalytical from the interaction of the analyte, organicor inorganic, with electromagnetic radiation indifferent wavelengths. They are dividedin atomic spectroscopic techniques and techniquesmolecular spectroscopic, where the firstobserve the effect of radiation absorption bya certain atom and the second observe theeffect of radiation absorption by a givenmolecule or chemical group.Currently, there are a large number of techniquesspectroscopic, highlighting those that havegreater application, such as: Spectrophotometry ofUV-Visible Absorption, SpectroscopyAbsorption in the Middle Infrared Regionwith Fourier Transform, SpectroscopyAtomic Absorption, Atomic Emission Spectroscopy, Fluorescence Spectroscopy andNuclear Magnetic Resonance.

Conclusion

Otto Gottlieb's legacy is beyond his research, he was an innovative researcher who proved he saw beyond his time and had an analytical eye for preserving species and maintaining biodiversity not only in his country but in the world.

References:

CAPARICA, Celira. Otto Gottlieb boosted a natural product chemistry. Cienc. Cult., Sao Paulo, v. 64, no. 1, p. 10-11, January 2012.

GOTTLIEB, O. R. 1989 . The role of oxygen in phytochemical evolution towards diversity. Phytochemistry.

vol. 28 , p. 2359 - 2362

GOTTLIEB, O. R. 1990 . Phytochemicals: differentiation and function. Phytochemistry. vol. 29 , p. 1715 -1724

GOTTLIEB, O. R. and BORIN, M. R. DE M. B. 1994 . The diversity of plants: Where is it? Why is it there?

What will it become. Anais da Academia Brasileira de Ciências. , 66, Sipl. 1, Part 1 . p. 55 - 83

GOTTLIEB, O. R. and BORIN, M. R. DE M. B. 1998 . Evolution of angiosperms via modulation of antagonisms. Phytochemistry. vol. 49 , p. 1 - 15

Publisher: Eliva Press SRL

Email: info@elivapress.com

Eliva Press is an independent publishing house established for the publication and dissemination of academic works all over the world. Company provides high quality and professional service for all of our authors.

Our Services:
Free of charge, open-minded, eco-friendly, innovational.

-Free standard publishing services (manuscript review, step-by-step book preparation, publication, distribution, and marketing).
-No financial risk. The author is not obliged to pay any hidden fees for publication.
-Editors. Dedicated editors will assist step by step through the projects.
-Money paid to the author for every book sold. Up to 50% royalties guaranteed.
-ISBN (International Standard Book Number). We assign a unique ISBN to every Eliva Press book.
-Digital archive storage. Books will be available online for a long time. We don't need to have a stock of our titles. No unsold copies. Eliva Press uses environment friendly print on demand technology that limits the needs of publishing business. We care about environment and share these principles with our customers.
-Cover design. Cover art is designed by a professional designer.
-Worldwide distribution. We continue expanding our distribution channels to make sure that all readers have access to our books.

www.elivapress.com

www.ingramcontent.com/pod-product-compliance
Lightning Source LLC
Chambersburg PA
CBHW051253170526
45165CB00004B/1695